PIONEERING TOMORROW'S AI SYSTEM THROUGH ELECTRICAL ENGINEERING

AN EMPIRICAL STUDY

OF THE PETER CHEW RULE

FOR OVERCOMING ERROR

IN CHAT GPT

CW00550973

PETER CHEW

PCET VENTURES (003368687-P)

Email:peterchew999@hotmail.my

© Peter Chew 2023

Cover Design : Peter Chew

Cover Image: Freepik Premium

No part of this book may be reproduced in any form or by any electronic or mechanical means, including information storage and retrieval systems, without written permission from the author.

Mathematician, Inventor and Biochemist Peter Chew

Peter Chew is Mathematician, Inventor and Biochemist. Global issue analyst, Reviewer for Europe Publisher, Engineering Mathematics Lecturer and President of Research and Development Secondary School (IND) for Kedah State Association [2015-18].

Peter Chew received the Certificate of appreciation from Malaysian Health Minister Datuk Seri Dr. Adam Baba(2021), PSB Singapore. National QC Convention STAR AWARD (2 STAR), 2019 Outstanding Analyst Award from IMRF International Multidisciplinary Research Foundation), IMFR Inventor Award 2020 , the Best Presentation Award at the 8th

International Conference on Engineering Mathematics and Physics ICEMP 2019 in Ningbo, China , Excellent award (Silver) of the virtual International, Invention, Innovation & Design Competition 2020 (3iDC2020) and Jury in the International Teaching and Learning Invention, Innovation Competition (iTaLiiC2023).

Analytical articles published in local and international media. Author for more than 60 Books , 8 preprint article published in the World Health Organization (WHO) , 4 article published in the Europe PMC and 22 are full article.

Peter Chew also is CEO PCET, Ventures, Malaysia, PCET is a long research associate of IMRF (International Multidisciplinary Research Foundation), Institute of higher Education & Research with its HQ at India and Academic Chapters all over the world, PCET also Conference Partner in CoSMEd2021 by SEAMEO RECSAM.

Peter Chew as 2nd Plenary Speaker the 6th International Multidisciplinary Research Conference with a Mindanao Zonal Assembly on January 14, 2023, at the Immaculate Conception University, Bajada Campus, Davao City.

Keynote Speaker of the 8th International Conference on Computer Engineering and Mathematical Sciences (ICCEMS 2019) , the International Conference on Applications of Physics , Chemistry & Engineering Sciences, ICPCE 2020 , 2nd Global Summit on Public Health and Preventive Medicine (GSPHPM2023) June 19, 2023 and World BIOPOLYMERS & POLYMER CHEMISTRY CONGRESS" 10-11 July 2023 | Online by Drug Delivery,

Special Talk Speaker at the 2019 International Conference on Advances in Mathematics, Statistics and Computer Science, the 100th CONF of the IMRF,2019, Goa , India.

Invite Speaker of the 24th Asian Mathematical Technology Conference (ATCM 2019) Leshan China , the 5th(2020), 6th (2021) and 7th (2022) International Conference on Management Engineering, Science, Social Sciences and Humanities by Society For Research Development(SRD) and 12th International Conference on Engineering Mathematics and Physics (July 5-7, 2023 in Kuala Lumpur, Malaysia).

Peter Chew is also Program Chair for the 11th International Conference on Engineering Mathematics and Physics (ICEMP 2022, Saint-Étienne, France | July 7-9, 2022) and Program Chair for the 12th International Conference on Engineering Mathematics and Physics (ICEMP 2023, Kuala Lumpur Malaysia | July 5-7, 2023).

For more information, please get it from this link Orcid https://orcid.org/0000-0002-5935-3041.

PIONEERING TOMORROW'S AI SYSTEM
THROUGH ELECTRICAL ENGINEERING
AN EMPIRICAL STUDY OF THE PETER CHEW METHOD
FOR OVERCOMING ERROR IN CHAT GPT

TABLE OF CONTENTS

PIONEERING TOMORROW'S AI SYSTEM THROUGH ELECTRICAL ENGINEERING AN EMPIRICAL STUDY OF THE PETER CHEW METHOD FOR OVERCOMING ERROR IN CHAT GPT

TABLE OF CONTENTS

Pioneering Tomorrow's AI System Through Electrical Engineering.

An Empirical Study Of The Peter Chew Method For Overcoming Error In Chat GPT

Abstract:

Introduction:

This empirical study investigates the Peter Chew Method for Overcoming Error In Chat GPT. – on enhancing Chat GPT's competence in effectively solving **Electrical Engineering** problem. The integration of Artificial Intelligence (AI) into Electrical Engineering problem -solving has paved the way for innovative approaches. This study aim to showcase the important of Peter Chew Method For Overcoming Error In AI system like GPT Chat.

Evidence:

Drawing upon empirical evidence, this study presents a comprehensive exposition of ChatGPT's adept utilization of Peter Chew Method correct solving Electrical Engineering problem that cannot be solved directly by the cosine and sine rules.

In stark contradistinction, the Method adopted by ChatGPT' can not correct solving **Electrical Engineering** problem tha cannot be solved directly by the cosine and sine rules. Thi underscores the pivotal role endowed by the Peter Chew Method in amplifying the solving **Electrical Engineering** problem proficiencies intrinsic to AI systems like Chat GPT.

Result :

The findings derived from this study unveil a compelling and notable demonstration of ChatGPT's adept utilization of the Peter Chew Method. This Method approach has yielded outcomes that are both substantial and convincing, particularly in the context of solving **Electrical Engineering** problem tha cannot be solved directly by the cosine and sine rules.

This study's results provide compelling evidence of ChatGPT' adept use of the Peter Chew Method, enabling correct solving **Electrical Engineering** problem that cannot be solved directly by the cosine and sine rules. In contrast, when ChatGPT using current approach, ChatGPT can not correct solving **Electrical Engineering** problem that cannot be solved directly by the cosine and sine rules.

his performance disparity underscores the vital role of the
'eter Chew Method in enhancing AI systems' solving
lectrical Engineering problem abilities, highlighting the
ransformative potential of diverse methodologies in advancing
\I's mathematical prowess.

Conclusion :

'ioneering Novel Maths Method e such as Peter Chew Method
or Solution of Triangle For Overcoming Errors in AI System
ike GPT Chat.

his study underscores the importance of pioneering innovative
Method to overcome existing Errors in AI systems like
ChatGPT, particularly in Solving Triangle Problem. The
groundbreaking Peter Chew Method for Solution of Triangle
howcased here holds the promise of unleashing untapped
ootential, elevating AI systems to new levels of proficiency.

Essentially, the Peter Chew Method offers a strategic avenue
or enhancing AI capabilities and pushing the boundaries of
ichievable accomplishments.

1

Discussion:

The outcomes derived from this study underscore the significant influence wielded by the method selection in augmenting the mathematical competencies of ChatGPT Particularly noteworthy is the application of the Peter Chew Method, which surfaces as a compelling exemplar. This Method serves as a overcomes current Errors on solving Electrical Engineering problem that cannot be solved directly by the cosine and sine rules in AI systems like ChatGPT.

Implications and Future Research:

These findings not only contribute to enhance AI's mathematical competencies but also emphasize the need for pioneering new Methods , Rules, Theorems, or Formulas to further enhance AI systems like ChatGPT. Future research could explore the development of novel mathematical techniques tailored to AI systems, thus expanding their capabilities across diverse problem-solving domains. This can be effective in let Electrical Engineering student interest in using *AI systems like ChatGPT* while learning Electrical Engineering especially when analogous covid- 19 issues arise in the future.

1.Background.

1.1 ChatGPT[1] :

Introducing ChatGPT.

We've trained a model called ChatGPT which interacts in a conversational way. The dialogue format makes it possible for ChatGPT to answer follow up questions, admit its mistakes, challenge incorrect premises, and reject inappropriate requests.

1.2 Introducing ChatGPT Plus

We're launching a pilot subscription plan for ChatGPT, a conversational AI that can chat with you, answer follow-up questions, and challenge incorrect assumptions. **Chat GPT[2]** knowledge is still limited to 2021 data, which means it can't answer current questions.

1.3 Knowledge is power: why the future is not just about the tech[3].

If we are to rely on machine intelligence, we need to understand the two types of knowledge. Understanding knowledge means we can distinguish where we want machines to do the mundane work and where we want humans to perform intuitive tasks.Such an approach will be as beneficial for business as for education.

As virtual and physical worlds become increasingly interdependent, knowledge – and how we manage it – will become the secret ingredient to manage the situation. And thrive. Virtual technologies are swiftly becoming intertwined with our physical world, and companies need to adapt. But that doesn't simply mean replacing humans with robots or relying on artificial intelligence (AI) to make all of our decisions.

This is because technology, though powerful, is just part of the equation. In fact, human intelligence will be one of the most valuable assets in today's Fourth Industrial Revolution (FIR), and companies may flounder if they fail to strike the right balance of automated technology and human insights.

1.4 Electrical **Engineering**[4].

Lesson 5. PHASE RELATIONS AND VECTOR REPRESENTATION

Parallelogram method

This technique is used for addition of two phasors at a time. The two alternating quantities are denoted by phasor diagram. The two phasors are arranged as the adjacent sides of a parallelogram. The diagonal of the formal parallelogram gives the resultant value of the two phasors. The following diagram shows phasor diagram of a.c parallel circuit:

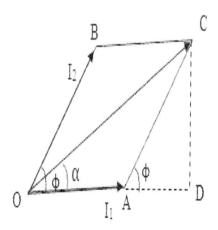

Fig 5.4 Phasor diagram of a.c parallel circuit

[A PHASOR DIAGRAM FOR AN AC CIRCUIT]

The two currents flowing in the circuit are given as:

$i_1 = I_{m1} Sin\ \omega t$

$i_2 = I_{m2}\ Sin\ (\omega t + \phi)$

i_r = resultant current

I_{m1} and I_{m2} are the maximum value of currents i_1 and i_2 respectively. Here i_1 is leading w.r.t i_2 or in other words i_2 i lagging w.r.t i_1. The phase difference between i_1 and i_2 is $\phi°$.

$$OC = \sqrt{(OD)^2 + (DC)^2}$$

$$= \sqrt{(OA + AD)^2 + (DC)^2}$$

$$= \sqrt{I_{m1}^2 + I_{m1}^2 (Sin^2\phi + Cos^2\phi) + 2I_{m1}I_{m2}Cos\phi}$$

$$= \sqrt{I_{m1}^2 + I_{m2}^2 + 2I_{m1}I_{m2}Cos\ \phi}$$

$$tan\ \alpha = \frac{y}{x} = \frac{CD}{OD}$$

$$Phase\ angle\ \alpha = tan^{-1}\frac{CD}{OD}$$

$$= tan^{-1}\frac{I_{m2}Sin\phi}{I_{m1} + I_{m2}Cos\phi}$$

The equation for instantaneous value of resultant current i_r i: given as: $i_r = I_{mr}\ Sin\ (\omega t + \alpha)$.

1.5   Electronics Tutorials.
Resultant Value of VT [5].

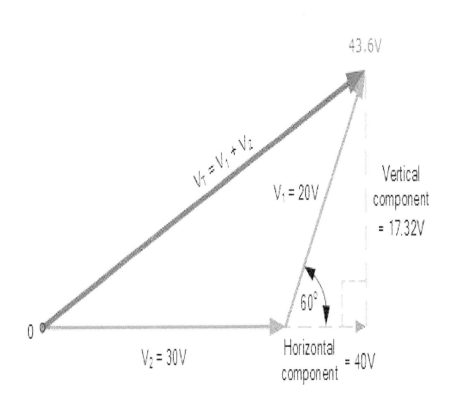

Phasor Subtraction of Phasor Diagrams

Phasor subtraction is very similar to the above rectangular method of addition, except this time the vector difference is the other diagonal of the parallelogram between the two voltages of V_1 and V_2 as shown.

2. Current Method and Peter Chew Method [6] for solution o
triangle.

Current Method and Peter Chew Method for solution o
triangle.

Example, Find third side, b .

Current Method 1:

Step 1

$$\frac{6}{\sin A} = \frac{4}{\sin 25°}$$

Rearranging gives us:

$$\sin A = \frac{6 \sin 25°}{4}$$

$$\sin A = 0.6339$$

$$A = 39.34° , 140.66°$$

Step 2:

When A = 39.34° , $B = 180° - 39.34° - 25° = 115.66°$

When A = 140.66°, $B = 180° - 140.66° - 25° = 14.34°$

Step 3:

When B = 115.66°

$$b^2 = 4^2 + 6^2 - 2(4)(6)\ cos115.66°$$
$$= 72.785$$
$$b = 8.531$$

When B = 14.34°

$$b^2 = 4^2 + 6^2 - 2(4)(6)\ cos14.34°$$
$$= 5.4955$$
$$b = 2.344$$

$\therefore$ b = 8.531 , 2.344

Current Method 2

Step 1

$$\frac{6}{\sin A} = \frac{4}{\sin 25°}$$

Rearranging gives us:

$$\sin A = \frac{6 \sin 25°}{4}$$

$$\sin A = 0.6339$$

$$A = 39.34° , 140.66$$

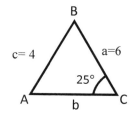

Step 2:

When A = 39.34°, $B = 180° - 39.34° - 25° = 115.66°$

When A = 140.66°, $B = 180° - 140.66° - 25° = 14.34°$

When B = 115.66°, $\dfrac{b}{\sin 115.66°} = \dfrac{4}{\sin 25°}$

Rearranging gives us:

$$b = \dfrac{4 \sin 115.66°}{\sin 25°}$$

$$= 8.531$$

When B = 14.34°, $\dfrac{b}{\sin 14.34°} = \dfrac{4}{\sin 25°}$

Rearranging gives us: $b = \dfrac{4 \sin 14.34°}{\sin 25°}$

$$= 2.344$$

∴ b = 8.531 , 2.344

Cosine rule:

$$4^2 = b^2 + 6^2 - 2(b)(6)\cos 25°$$

$$b^2 - 10.8757\ b + 20 = 0$$

$$b = 8.531 , 2.344$$

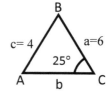

3. Using Math portal for solving triangle problem [Calculated on Feb 1, 2023]

3.1 MathPortal website's[7] owner is mathematician Miloš Petrović.

MathPortal use current method 1.

Example: Find side a of a triangle if side b=3, side c = 5 and angle C = 35°

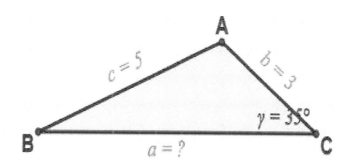

explanation

STEP 1: find angle β

To find angle β use Thw Law of Sines:

$$\frac{\sin(\beta)}{b} = \frac{\sin(\gamma)}{c}$$

After substituting $b = 3$, $c = 5$ and $\gamma = 35°$ we have:

$$\frac{\sin(\beta)}{3} = \frac{\sin(35°)}{5}$$

$$\frac{\sin(\beta)}{3} = \frac{0.5736}{5}$$

$$\sin(\beta) \cdot 5 = 3 \cdot 0.5736$$

$$\sin(\beta) \cdot 5 = 1.7207$$

$$\sin(\beta) = \frac{1.7207}{5}$$

$$\sin(\beta) = 0.3441$$

$$\beta = \arcsin(0.3441)$$

$$\beta \approx 20.1297^{o}$$

STEP 2: find angle α

To find angle α use formula:

$$\alpha + \beta + \gamma = 180^{o}$$

After substituting $\beta = 20.1297^{o}$ and $\gamma = 35^{o}$ we have:

$$\alpha + 20.1297^{o} + 35^{o} = 180^{o}$$

$$\alpha + 55.1297^{o} = 180^{o}$$

$$\alpha = 180^{o} - 55.1297^{o}$$

$$\alpha = 124.8703^{o}$$

STEP 3: find side a

To find side a use Law of Cosines:

$$a^2 = b^2 + c^2 - 2 \cdot b \cdot c \cdot \cos(\alpha)$$

After substituting $b = 3$, $c = 5$ and $\alpha = 124.8703°$ we have:

$$a^2 = 3^2 + 5^2 - 2 \cdot 3 \cdot 5 \cdot \cos(124.8703°)$$

$$a^2 = 9 + 25 - 2 \cdot 3 \cdot 5 \cdot \cos(124.8703°)$$

$$a^2 = 34 - 2 \cdot 15 \cdot \cos(124.8703°)$$

$$a^2 = 34 - 30 \cdot (-0.5717)$$

$$a^2 = 34 - (-17.1516)$$

$$a^2 = 51.1516$$

$$a = \sqrt{51.1516}$$

$$a \approx 7.152$$

3.2 Peter Chew Method

Using cosine rule:

$$5^2 = a^2 + 3^2 - 2(a)(3)\,cos35°$$

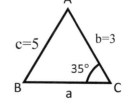

$$a^2 - 4.9149\ a - 16 = 0$$

$$\boldsymbol{a} = 7.152,\ -2.237\ \text{(rejected, a > 0)}$$

4. Pioneering Tomorrow's AI System Through Electrical Engineering

4.1   Electrical Engineering problems that can be solved directly by the Cosines Rule

Example 4.1 A phasor diagram for an AC circuit

can be show in triangle ABC.

If V_1 =AB = 20 V , V_2 = BC = 80 V   ?

and angle ABC= 120°. Determine

the resultant(AC) .

C

$V_2 = 80V$

120°

A $V_1 = 20V$ B

Current Solution

$\angle$ OBA $= 180° - 60° = 120°$

Applying the cosine rule,

$$AC^2 = V_1^2 + V_2^2 - 2\ V_1\ V_2 \cos \angle ABC$$

$$= 20^2 + 80^2 - 2\ (20)(80) \cos 120°$$

$$= 8\ 400$$

The resultant, AC $= \sqrt{8\ 400} = 91.65$ **V**

Chat **GPT**[8] can correctly solve **Electrical** <u>**Engineering**</u> problems that can be solved directly by the Cosines Rule.

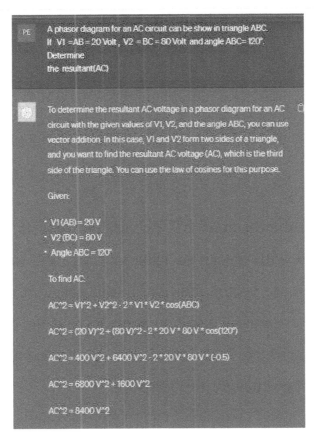

PE A phasor diagram for an AC circuit can be show in triangle ABC. If V1 =AB = 20 Volt , V2 = BC = 80 Volt and angle ABC = 120°. Determine the resultant(AC)

To determine the resultant AC voltage in a phasor diagram for an AC circuit with the given values of V1, V2, and the angle ABC, you can use vector addition. In this case, V1 and V2 form two sides of a triangle, and you want to find the resultant AC voltage (AC), which is the third side of the triangle. You can use the law of cosines for this purpose.

Given:

- V1 (AB) = 20 V
- V2 (BC) = 80 V
- Angle ABC = 120°

To find AC:

AC^2 = V1^2 + V2^2 - 2 * V1 * V2 * cos(ABC)

AC^2 = (20 V)^2 + (80 V)^2 - 2 * 20 V * 80 V * cos(120°)

AC^2 = 400 V^2 + 6400 V^2 - 2 * 20 V * 80 V * (-0.5)

AC^2 = 6800 V^2 + 1600 V^2

AC^2 = 8400 V^2

Now, take the square root of both sides to find AC:

AC = √(8400 V^2)

AC ≈ 91.65 V

So, the resultant AC voltage (AC) is approximately 91.65 volts.

Full text:

A phasor diagram for an AC circuit can be show in triangle ABC. If V1 =AB = 20 Volt , V2 = BC = 80 Volt and angle ABC= 120°. Determine the resultant(AC)

ChatGPT

To determine the resultant AC voltage in a phasor diagram for an AC circuit with the given values of V1, V2, and the angle ABC, you can use vector addition. In this case, V1 and V2 form two sides of a triangle, and you want to find the resultant AC voltage (AC), which is the third side of the triangle. You can use the law of cosines for this purpose.

Given: V1 (AB) = 20 V

V2 (BC) = 80 V

Angle ABC = 120°

o find AC:

$AC^2 = V1^2 + V2^2 - 2 * V1 * V2 * cos(ABC)$

$AC^2 = (20 V)^2 + (80 V)^2 - 2 * 20 V * 80 V * cos(120°)$

$AC^2 = 400 V^2 + 6400 V^2 - 2 * 20 V * 80 V * (-0.5)$

$AC^2 = 6800 V^2 + 1600 V^2$

$AC^2 = 8400 V^2$

Now, take the square root of both sides to find AC:

$AC = \sqrt{(8400 V^2)}$

$AC \approx 91.65 V$

So, the resultant AC voltage (AC) is approximately 91.65 volts.

4.2 Electrical Engineering problems that can be solved directly by the Sine Rule .

Chat GPT can correctly solve Electrical Engineering problems that can be solved directly by the Sine Rule

Example 4.2 A phasor diagram for an AC circuit can be show in triangle ABC.

If $V_2 = BC = 80$ V , the resultant(AC) $= 91.65$ V and angle ABC= 120°.

$V_r = 91.65V$

$120°$ $V_2 = 80V$

?

A   B

C

Determine the angle CAB [angle of the resultant(AC) makes with V_1 (AB)].

Sine rule:

$$\frac{91.65}{\sin 120°} = \frac{80}{\sin \angle CAB}$$

$$\sin \angle CAB = \frac{80 \; \sin 120°}{91.65}$$

$$= 0.75594$$

$\angle CAB = 49.108°, 130.9°$ (which is impossible in this case)

Hence the angle the resultant makes with V_1 is 49.108° .

Chat **GPT** can correctly solve **Electrical** Engineering problems that can be solved directly by the Sine Rule.

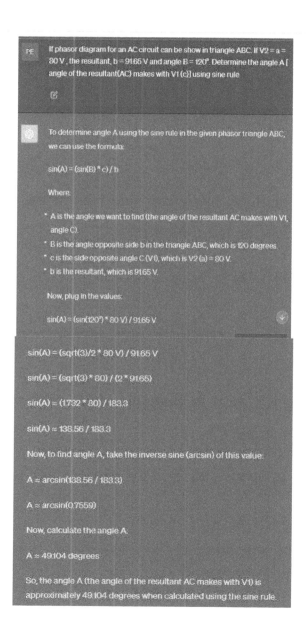

3

Full Text: If phasor diagram for an AC circuit can be show ir triangle ABC. If V2 = a = 80 V , the resultant, b = 91.65 V anc angle B = 120°. Determine the angle A [angle of the resultant(AC) makes with V1 (c)] using sine rule

ChatGPT

To determine angle A using the sine rule in the given phasor triangle ABC, we can use the formula:

$\sin(A) = (\sin(B) * c) / b$

Where:

A is the angle we want to find (the angle of the resultant AC makes with V1, angle C).

B is the angle opposite side b in the triangle ABC, which is 120 degrees.

c is the side opposite angle C (V1), which is V2 (a) = 80 V.

b is the resultant, which is 91.65 V.

Now, plug in the values:

$\sin(A) = (\sin(120°) * 80 \text{ V}) / 91.65 \text{ V}$

$\sin(A) = (\text{sqrt}(3)/2 * 80 \text{ V}) / 91.65 \text{ V}$

$\sin(A) = (\text{sqrt}(3) * 80) / (2 * 91.65)$

$\sin(A) \approx (1.732 * 80) / 183.3$

$\sin(A) \approx 138.56 / 183.3$

Now, to find angle A, take the inverse sine (arcsin) of this value:

$A \approx \arcsin(138.56 / 183.3)$

$A \approx \arcsin(0.7559)$

Now, calculate the angle A:

$A \approx 49.104$ degrees

So, the angle A (the angle of the resultant AC makes with V1) is approximately 49.104 degrees when calculated using the sine rule.

4.3 Electrical Engineering problems that can not be solved directly by the Cosine and Sine Rule.

I) Chat GPT cannot correctly solve Civil Engineering problems that can not be solved directly by the Cosine and Sine Rule.

II) Chat GPT uses Peter Chew's Method to correctly solve Civil Engineering problems that cannot be solved directly by cosine and sine rules .

Example 4.3.1 : A phasor diagram for an AC circuit can be show in triangle ABC. If $V_{AB} = 50$ V , $V_{BC} = 30$ V and angle $C = 20°$. Determine the resultant(V_{AC}) .

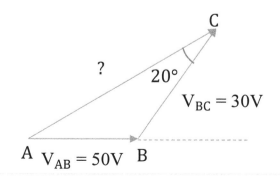

Current Method 1:

Step 1

$$\frac{30}{\sin A} = \frac{50}{\sin 20°}$$

Rearranging gives us:

$$\sin A = \frac{30 \sin 20°}{50}$$

$$\sin A = 0.2052$$

$$A = 11.84° , 168.16° \text{ (Reject, because } 168.16°+20° > 180°)$$

Step 2:

When A = 11.84° , $B = 180° - 11.84° - 20° = 148.16°$

Step 3:

When B = 148.16°

$$b^2 = 50^2 + 30^2 - 2(50)(30) \cos 148.16°$$

$$= 5\,949$$

$$b = 77.13$$

∴ $V_{AC} = 77.13$ V

Step 1

$$\frac{30}{\sin A} = \frac{50}{\sin 20°}$$

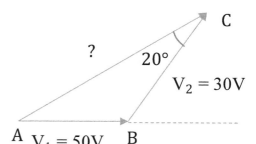

? 20° C

$V_2 = 30V$

A $V_1 = 50V$ B

Rearranging gives us:

$$\sin A = \frac{30 \sin 20°}{50}$$

$$\sin A = 0.2052$$

$A = 11.84°$, $168.16°$ (Reject, because $168.16°+20° > 180°$)

Step 2:

When A = $11.84°$, $B = 180° - 11.84° - 20° = 148.16°$

When B = $148.16°$, $\dfrac{b}{\sin 148.16°} = \dfrac{50}{\sin 20°}$

Rearranging gives us: $b = \dfrac{50 \sin 148.16°}{\sin 20°}$

$$= 77.12$$

$\therefore$ $V_{AC} = 77.13$ V

Peter Chew Method

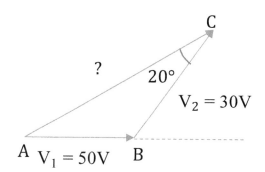

Cosine rule:

$$50^2 = b^2 + 30^2 - 2(b)(30)\ cos20°$$

$$b^2 - 56.38\ b - 1600 = 0$$

$b = 77.13$, -20.75(Reject because b>0)

$\therefore\ \ V_{AC} = 77.13$ V

4.3.1 I) Chat GPT cannot correctly solve Electrica Engineering problems that can not be solved directly by the Cosine and Sine Rule .

Example 4.3.1 *If* phasor diagram for an AC circuit can be show in triangle ABC(ABC is a non-right-angled triangle). I V AB or c = 50 V and V BC or a = 30 V and angle C = 20° Determine the resultant(AC) or b in decimal value.

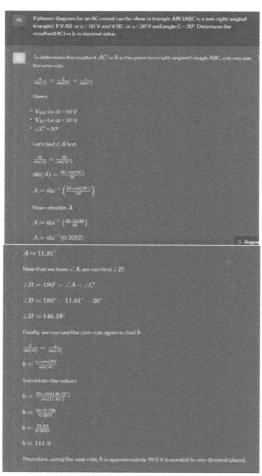

Full text. Bold Text: Error Step

If phasor diagram for an AC circuit can be show in triangle ABC(ABC is a non-right-angled triangle). If V AB or c = 50 V and V BC or a = 30 V and angle C = 20°. Determine the resultant(AC) or b in decimal value.

ChatGPT

To determine the resultant AC or b in the given non-right-angled triangle ABC, you can use the sine rule: $\frac{a}{\sin A} = \frac{b}{\sin B} = \frac{c}{\sin C}$

Given: VAB (or c) = 50 V, VBC (or a) = 30 V, $\angle C = 20°$

Let's find $\angle A$ first:

$$\frac{30}{\sin A} = \frac{50}{\sin 20°}$$

$$\sin(A) = \frac{30 \sin 20°}{50}$$

$$A = \sin-1\left(\frac{30 \sin 20°}{50}\right)$$

Now calculate A:

$$A \approx \sin-1\left(\frac{30 \times 0.342}{50}\right)$$

$$A \approx \sin-1(0.2052)$$

$$A \approx 11.81°$$

8

Now that we have $\angle A$, we can find $\angle B$:

$\angle B = 180° - \angle A - \angle C$

$\angle B = 180° - 11.81° - 20°$

$\angle B \approx 148.19°$

Finally, we can use the sine rule again to find b:

$$\frac{b}{\sin B} = \frac{a}{\sin A}$$

$$b = \frac{a \sin B}{\sin A}$$

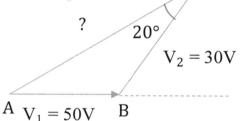

Substitute the values:

$$b = \frac{30 \times \sin(148.19°)}{\sin(11.81°)}$$

$$b = \frac{30 \times 0.766}{0.2052}$$

[Note: $\sin(148.19°) = 0.5271$, $\sin(11.81°) = 0.2047$]

$$b = \frac{22.98}{0.2052}$$

$b \approx 111.9$

Therefore, using the sine rule, b is approximately 111.9 V (rounded to one decimal place).

4.3.1. II)   Chat GPT uses Peter Chew's Method to correctly solve **Electrical Engineering** problems that cannot be solved directly by cosine and sine rules .

4.3.1 If phasor diagram for an AC circuit can be show in triangle ABC. If V1 or c = 50 V and V2 or a = 30 V and angle C = 20°. Determine the resultant(AC) or b using Peter Chew Method for Solution of triangle. For Peter Chew Method , first, direct substitute value to b^2 – (2a cos C) b + a^2 – c^2 =0, then solve the quadratic equation b^2 – (2a cos C) b + a^2 – c^2 =0 to find value of b in decimal .

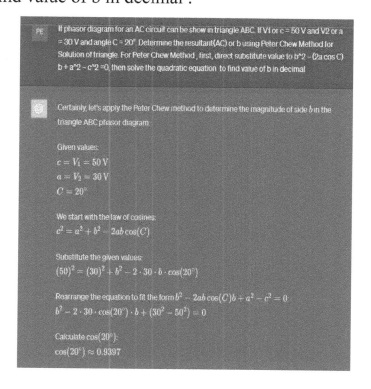

Now, substitute the values into the quadratic equation:

$$b^2 - 2 \cdot 30 \cdot 0.9397 \cdot b + (30^2 - 50^2) = 0$$

This is a quadratic equation in the form $Ax^2 + Bx + C = 0$, where $A = 1, B = -2 \cdot 30 \cdot 0.9397$, and $C = 30^2 - 50^2$.

Use the quadratic formula to solve for b:

$$b = \frac{-B \pm \sqrt{B^2 - 4AC}}{2A}$$

Substitute the values for A, B, and C into the quadratic formula and simplify:

$$b = \frac{56.382 \pm \sqrt{9570}}{2}$$

$$b = \frac{56.382 \pm 97.868}{2}$$

Now, there are two possible solutions:

$$b_1 = \frac{56.382 + 97.868}{2} \approx 77.125$$
$$b_2 = \frac{56.382 - 97.868}{2} \approx -20.743$$

Since length cannot be negative, we discard the negative solution. Therefore, $b \approx 77.125$.

So, using the Peter Chew method, the value of b is approximately 77.125. This represents the magnitude of the phasor corresponding to side BC in the phasor diagram triangle ABC.

**Note: Chat GPT can use Peter Chew's Method to correctl
solve Electrical Engineering   problems that cannot b
solved directly by cosine and sine rules because the solutio
becomes simple and involves fewer steps.**

Full Text: If phasor diagram for an AC circuit can be show in triangle ABC. If V1 or c = 50 V and V2 or a = 30 V and angle C = 20°. Determine the resultant(AC) or b using Peter Chew Method for Solution of triangle. For Peter Chew Method , first, direct substitute value to b^2 − (2a cos C) b + a^2 − c^2 =0, then solve the quadratic equation to find value of b in decimal

ChatGPT: Certainly, let's apply the Peter Chew method to determine the magnitude of side b in the triangle ABC phasor diagram.

Given values: $c=V1$ =50V ,$a=V2$ =30V ,$C=20∘$

We start with the law of cosines: $c2=a2+b2−2abcos(C)$

Substitute the given values:

$(50)2=(30)2+b2−2·30·b·cos(20∘)$

Rearrange the equation to fit the form

$b2−2abcos(C)b+a2−c2=0$:

$b2−2·30·cos(20∘)·b+(302−502)=0$

Calculate $cos(20∘)$:

$cos(20∘)≈0.9397$

Now, substitute the values into the quadratic equation:

$b2-2\cdot30\cdot0.9397\cdot b+(302-502)=0$

This is a quadratic equation in the form $Ax2+Bx+C=0$,

where $A=1$, $B=-2\cdot30\cdot0.9397$, and $C=302-502$.

Use the quadratic formula to solve for $b=\dfrac{-B\pm\sqrt{B^2-4AC}}{2A}$

$b=\dfrac{-B\pm\sqrt{B^2-4AC}}{2A}$

Substitute the values for A, B, and C into the quadratic formula and simplify:

$b=\dfrac{56.382\pm\sqrt{9578.793924}}{2}$

$b=\dfrac{56.382\pm97.878}{2}$

Now, there are two possible solutions:

$b1=\dfrac{56.382\pm97.878}{2}\approx77.130$

$b2=\dfrac{56.382-97.878}{2}\approx-20.748$

Since length cannot be negative, we discard the negative solution. Therefore, $b\approx77.125$.

So, using the Peter Chew method, the value of b is approximately 77.125. This represents the magnitude of the phasor corresponding to side *BC* in the phasor diagram triangle ABC.

Example 4.3.2 : A phasor diagram for an AC circuit can be show in triangle ABC. If $V_{AB} = 72$ V , $V_{BC} = 38$ V and angle C $= 50°$. Determine the resultant(V_{AC}) .

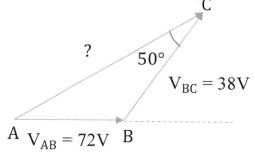

Current Method 1:

Step 1

$$\frac{38}{\sin A} = \frac{72}{\sin 50°}$$

Rearranging gives us:

$$\sin A = \frac{38 \sin 50°}{72}$$

$$\sin A = 0.4043$$

A $=$ 23.85° , 156.15° (Reject, because 156.15°+50° > 180°)

Step 2:

When A = 23.85° , B = 180° - 23.85° - 50° = 106.15°

Step 3:

When B = 106.15°

$$b^2 = 72^2 + 38^2 - 2(72)(38)\,cos106.15°$$

$$= 8\,150$$

$$b = 90.28$$

∴ V_{AC} = 90.28 V

Current Method 2

Step 1

$$\frac{38}{\sin A} = \frac{72}{\sin 50°}$$

Rearranging gives us:

$$\sin A = \frac{38\,\sin 50°}{72}$$

$$\sin A = 0.4043$$

A = 23.85° , 156.15° (Reject, because 156.15°+50° > 180°)

Step 2:

When A = 23.85° , B = 180° - 23.85° - 50° = 106.15°

When B = 106.15° , $\dfrac{b}{\sin 106.15°} = \dfrac{72}{\sin 50°}$

Rearranging gives us: b $= \dfrac{72 \sin 106.15°}{\sin 50°}$

$\qquad\qquad\qquad = \ 90.28$

∴ $\quad V_{AC} = 90.28$ V

Peter Chew Method

Cosine rule:

$\qquad 72^2 = b^2 + 38^2 - 2(b)(38) \ cos50°$

$\qquad b^2 - 48.85 \ b - 3740 = 0$

$\qquad b = 90.28 \ , \ -41.43 (\text{Reject because b>0})$

∴ $\ V_{AC} = 90.28$ V

4.3.2 I)   Chat GPT cannot   correctly solve   Electrica

Engineering problems that can not   be solved directly by the

Cosine and Sine Rule .

4.3.2   If phasor diagram for an AC circuit can be show i

triangle ABC(ABC is a non-right-angled triangle). If V AB or

= 72 V and V BC or a = 38 V and angle C = 50°. Determine th

resultant(AC) or b in decimal value.

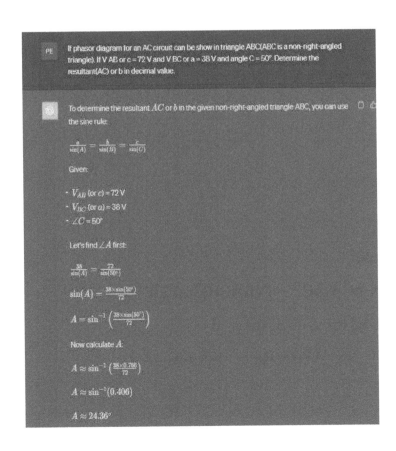

Now that we have $\angle A$, we can find $\angle B$:

$$\angle B = 180^\circ - \angle A - \angle C$$

$$\angle B = 180^\circ - 24.36^\circ - 50^\circ$$

$$\angle B \approx 105.64^\circ$$

Finally, we can use the sine rule again to find b:

$$\frac{b}{\sin(B)} = \frac{a}{\sin(A)}$$

$$b = \frac{a \times \sin(B)}{\sin(A)}$$

Substitute the values:

$$b = \frac{38 \times \sin(105.64^\circ)}{\sin(24.36^\circ)}$$

$$b \approx \frac{38 \times 0.573}{0.406}$$

$$b \approx \frac{21.374}{0.406}$$

$$b \approx 53.75$$

Therefore, using the sine rule, b is approximately 53.75 V (rounded to two decimal places).

Full text. Bold Text: Error Step

f phasor diagram for an AC circuit can be show in triangle ABC(ABC is a non-right-angled triangle). If V AB or c = 72 V nd V BC or a = 38 V and angle C = 50°. Determine the esultant(AC) or b in decimal value.

ChatGPT

To determine the resultant AC or b in the given non-right-angled triangle ABC, you can use the sine rule:

$$\frac{a}{\sin A} = \frac{b}{\sin B} = \frac{c}{\sin C}$$

Given:

VAB   (or c) = 72 V

VBC   (or a) = 38 V

$\angle C = 50°$

Let's find $\angle A$ first:

$$\frac{38}{\sin A} = \frac{72}{\sin 50°}$$

$\sin(A) = \frac{38 \sin 50°}{72}$

$A = \sin-1(\frac{38 \sin 50°}{72})$

Now calculate A:

$A \approx \sin-1(\frac{38 \times 0.766}{72})$

$A \approx \sin-1(0.406)$ [Note: $\frac{38 \times 0.766}{72} = 0.4043$]

$A \approx 24.36\circ$

Now that we have $\angle A$, we can find $\angle B$:

$\angle B = 180\circ - \angle A - \angle C$

$\angle B = 180\circ - 24.36\circ - 50\circ$

$\angle B \approx 105.64\circ$

Finally, we can use the sine rule again to find b:

$$\frac{b}{\sin B} = \frac{a}{\sin A}$$

$$b = \frac{a \sin B}{\sin A}$$

Substitute the values:

$$b = \frac{38 \times \sin(105.64°)}{\sin(24.36°)}$$

$$b = \frac{38 \times 0.573}{0.406}$$

[Note: $\sin(105.64°) = 0.9630$, $\sin(24.36°) = 0.4125$]

$$b \approx \frac{21.834}{0.406}$$

$$b \approx 53.75$$

Therefore, using the sine rule, b is approximately 53.75 V rounded to two decimal places).

4.3.2 II) Chat GPT uses Peter Chew's Method to correctly solve **Electrical Engineering** problems that cannot be solved directly by cosine and sine rules .

4.3.2 If phasor diagram for an AC circuit can be show in triangle ABC. If V1 or c = 72 V and V2 or a = 38V and angle C = 50°. Determine the resultant(AC) or a using Peter Chew Method for Solution of triangle. For Peter Chew Method , first, direct substitute value to b^2 – (2a cos C) b + a^2 – c^2 =0, then solve the quadratic equation b^2 – (2a cos C) b + a^2 – c^2 =0 to find value of b in decimal .

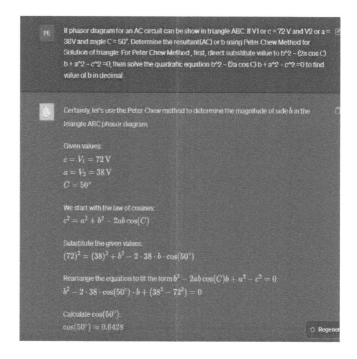

Now, substitute the values into the quadratic equation:

$$b^2 - 2 \cdot 38 \cdot 0.6428 \cdot b + (38^2 - 72^2) = 0$$

This is a quadratic equation in the form $Ax^2 + Bx + C = 0$, where $A = 1$, $B = -2 \cdot 38 \cdot 0.6428$, and $C = 38^2 - 72^2$.

Use the quadratic formula to solve for b:

$$b = \frac{-B \pm \sqrt{B^2 - 4AC}}{2A}$$

Substitute the values for A, B, and C into the quadratic formula and simplify:

$$b = \frac{48.8528 \pm \sqrt{17347}}{2}$$

$$b = \frac{48.8528 + 131.6876}{2}$$

Now, there are two possible solutions:

$$b_1 = \frac{48.8528 + 131.6876}{2} \approx 90.2702$$

$$b_2 = \frac{48.8528 - 131.6876}{2} \approx -41.4174$$

Since length cannot be negative, we discard the negative solution. Therefore, $b \approx 90.2702$.

So, using the Peter Chew method, the value of b is approximately 90.2702. This represents the magnitude of the phasor corresponding to side BC in the phasor diagram triangle ABC.

Note: Chat GPT can use Peter Chew's Method to correctly solve Electrical Engineering problems that cannot be solved directly by cosine and sine rules because the solution becomes simple and involves fewer steps.

Full text: **If phasor diagram for an AC circuit can be show i triangle ABC. If V1 or c = 72 V and V2 or a = 38V and angle (= 50°. Determine the resultant(AC) or b using Peter Chew Method for Solution of triangle. For Peter Chew Method , firs direct substitute value to b^2 – (2a cos C) b + a^2 – c^2 =(then solve the quadratic equation b^2 – (2a cos C) b + a^2 – c^ =0 to find value of b in decimal .**

ChatGPT

Certainly, let's use the Peter Chew method to determine the magnitude of side b in the triangle ABC phasor diagram.

Given values: $c = V1$ $= 72V$ $a = V2$ $= 38V$ $C = 50\circ$

We start with the law of cosines: $c2 = a2 + b2 - 2ab\cos(C)$

Substitute the given values:

$(72)2 = (38)2 + b2 - 2 \cdot 38 \cdot b \cdot \cos(50\circ)$

Rearrange the equation to fit the form

$b2 - 2ab\cos(C)b + a2 - c2 = 0$:

$b2 - 2 \cdot 38 \cdot \cos(50\circ) \cdot b + (382 - 722) = 0$

Calculate $\cos(50\circ)$:

$\cos(50\circ) \approx 0.6428$

Now, substitute the values into the quadratic equation:

$b2-2\cdot38\cdot0.6428\cdot b+(382-722)=0$

This is a quadratic equation in the form $Ax2+Bx+C=0$, where $A=1$, $B=-2\cdot38\cdot0.6428$, and $C=382-722$.

Use the quadratic formula to solve for $b = \dfrac{-B\pm\sqrt{B^2-4AC}}{2A}$

$$=\dfrac{-B\pm\sqrt{B^2-4AC}}{2A}$$

Substitute the values for A, B, and C into the quadratic formula and simplify:

$$=\dfrac{48.8528\pm\sqrt{17347}}{2}$$

$$=\dfrac{48.8528\pm131.6876}{2}$$

Now, there are two possible solutions:

$$b1 = \dfrac{48.8528+131.6876}{2} \qquad \approx 90.2702$$

$$b2 = \dfrac{48.8528-131.6876}{2} \approx -41.4174$$

Since length cannot be negative, we discard the negative solution. Therefore, $b\approx90.2702$.

So, using the Peter Chew method, the value of b is approximately 90.2702. This represents the magnitude of the

phasor corresponding to side BC in the phasor diagram triang
ABC.

Example 4.3.3 : A phasor diagram for an AC circuit can t
show in triangle ABC. If $V_{AB} = 85$ V , $V_{BC} = 26$ V and ang
$C = 35°$. Determine the resultant(V_{AC}) .

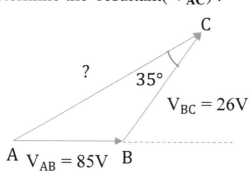

Current Method 1:

Step 1

$$\frac{26}{\sin A} = \frac{85}{\sin 35°}$$

Rearranging gives us:

$$\sin A = \frac{26 \sin 35°}{85}$$

$$\sin A = 0.1754$$

$$A = 10.10° , 169.90° \text{ (Reject, because } 169.90°+35° > 180°}$$

Step 2:

When A = 10.10° , $B = 180° - 10.10° - 35° = 134.9°$

Step 3:

When B = 134.9°

$$b^2 = 85^2 + 26^2 - 2(85)(26)\ cos134.9°$$

$$= 11021$$

$$b = 104.98$$

$\therefore\ V_{AC} = 104.98\ V$

Current Method 2

Step 1

$$\frac{26}{sin\ A} = \frac{85}{sin\ 35°}$$

Rearranging gives us:

$$sin\ A = \frac{26\ sin\ 35°}{85}$$

$$sin\ A = 0.1754$$

$A = 10.10° , 169.90°$ (Reject, because 169.90°+35° > 180°)

5

Step 2:

When A = 10.10° , $B = 180° - 10.10° - 35° = 134.9°$

When B = 134.9° , $\qquad \dfrac{b}{\sin 134.9°} = \dfrac{85}{\sin 35°}$

Rearranging gives us: b $= \dfrac{85 \ \sin 134.9°}{\sin 35°}$

$$= 104.97$$

$\therefore \quad V_{AC} = 104.97 \text{ V}$

Peter Chew Method

Cosine rule:

$$85^2 = b^2 + 26^2 - 2(b)(26) \ cos35°$$

$$b^2 - 42.596 \ b - 6549 = 0$$

$$b = 104.98 \ , \ - 62.384 \text{(Reject because b>0)}$$

$\therefore \quad V_{AC} = 104.98 \ \text{V}$

4.3.3 I)   Chat GPT cannot   correctly solve   Electrical
Engineering problems that can not   be solved directly by the
Cosine and Sine Rule .

4.3.3 If phasor diagram for an AC circuit can be show in
triangle ABC(ABC is a non-right-angled triangle). If V AB or c
= 85 V and V BC or a = 26 V and angle C = 35°. Determine the
resultant(AC) or b in decimal value.

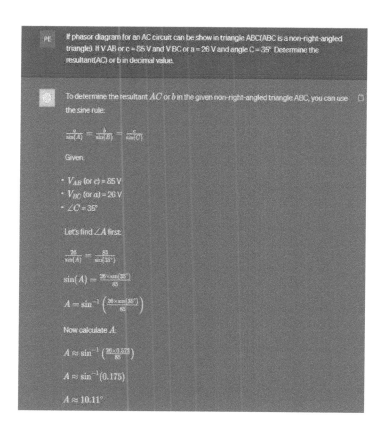

Now that we have $\angle A$, we can find $\angle B$:

$\angle B = 180° - \angle A - \angle C$

$\angle B = 180° - 10.11° - 35°$

$\angle B \approx 134.89°$

Finally, we can use the sine rule again to find b:

$\frac{b}{\sin(B)} = \frac{a}{\sin(A)}$

$b = \frac{a \times \sin(B)}{\sin(A)}$

Substitute the values:

$b = \frac{26 \times \sin(134.89°)}{\sin(10.11°)}$

$b \approx \frac{26 \times 0.766}{0.175}$

$b \approx \frac{19.92}{0.175}$

$b \approx 114.06$

Therefore, using the sine rule, b is approximately 114.06 V (rounded to two decimal places).

Note: $\sin 134.9° ==0.7085$, not 0.766 (Chat GPT gives wrong steps), therefore, the final answer is also wrong (114.06).

4.3.3 II) Chat GPT uses Peter Chew's Method to correctly solve **Electrical Engineering** problems that cannot be solved directly by cosine and sine rules .

4.3.3 If phasor diagram for an AC circuit can be show in triangle ABC. If V1 or c = 85 V and V2 or a = 26V and angle C = 35°. Determine the resultant(AC) or b using Peter Chew Method for Solution of triangle. For Peter Chew Method , first, direct substitute value to b^2 − (2a cos C) b + a^2 − c^2 =0, then solve the quadratic equation b^2 − (2a cos C) b + a^2 − c^2 =0 to find value of b in decimal

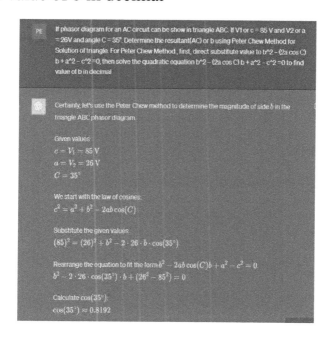

Now, substitute the values into the quadratic equation:

$$b^2 - 2 \cdot 26 \cdot 0.8192 \cdot b + (26^2 - 85^2) = 0$$

6

This is a quadratic equation in the form $Ax^2 + Bx + C = 0$, where $A = 1, B = -2 \cdot 26 \cdot 0.8192$, and $C = 26^2 - 85^2$.

Use the quadratic formula to solve for b:

$$b = \frac{-B \pm \sqrt{B^2 - 4AC}}{2A}$$

Substitute the values for A, B, and C into the quadratic formula and simplify:

$$b = \frac{42.5984 \pm \sqrt{28011}}{2}$$

$$b = \frac{42.5984 \pm 167.36}{2}$$

Now, there are two possible solutions:

$$b_1 = \frac{42.5984 + 167.36}{2} \approx 104.979$$
$$b_2 = \frac{42.5984 - 167.36}{2} \approx -62.381$$

Since length cannot be negative, we discard the negative solution. Therefore, $b \approx 104.979$.

So, using the Peter Chew method, the value of b is approximately 104.979. This represents the magnitude of the phasor corresponding to side BC in the phasor diagram triangle ABC.

Note: Chat GPT can use Peter Chew's Method to correctly solve Electrical Engineering problems that cannot be solved directly by cosine and sine rules because the solution becomes simple and involves fewer steps.

Full Text: If phasor diagram for an AC circuit can be show in triangle ABC. If V1 or c = 85 V and V2 or a = 26V and angle C = 35°. Determine the resultant(AC) or b using Peter Chew Method for Solution of triangle. For Peter Chew Method , first, direct substitute value to b^2 − (2a cos C) b + a^2 − c^2 =0, then solve the quadratic equation b^2 − (2a cos C) b + a^2 − c^2 =0 to find value of b in decimal .

ChatGPT:

Certainly, let's use the Peter Chew method to determine the magnitude of side b in the triangle ABC phasor diagram..

Given values $Vc=V1$ =85V $a=V2$ =26V $C=35\circ$

We start with the law of cosines: $c2=a2+b2-2abcos(C)$

Substitute the given values: $(85)2=(26)2+b2-2\cdot26\cdot b\cdot cos(35\circ)$

Rearrange the equation to fit the form $b2-2abcos(C)b+a2-c2=0$:

$b2-2\cdot26\cdot cos(35\circ)\cdot b+(262-852)=0$

Calculate $cos(35\circ)$:

$cos(35\circ)\approx0.8192$

Now, substitute the values into the quadratic equation:

$b2-2\cdot26\cdot0.8192\cdot b+(262-852)=0$

This is a quadratic equation in the form $Ax^2+Bx+C=0$, where $A=1$, $B=-2\cdot26\cdot0.8192$, and $C=262-852$.

Use the quadratic formula to solve for $b=\dfrac{-B\pm\sqrt{B^2-4AC}}{2A}$

Substitute the values for A, B, and C into the quadratic formula and simplify:

$$b=\frac{42.5984\pm\sqrt{28011}}{2}$$

$$b=\frac{42.5984\pm167.36}{2}$$

Now, there are two possible solutions:

$$b1 = \frac{42.5984+167.36}{2}\approx104.979$$

$$b2 = \frac{42.5984-167.36}{2}\approx-62.381$$

Since length cannot be negative, we discard the negative solution.

Therefore, $b\approx104.979$.

So, using the Peter Chew Method, the value of b is approximately 104.979. This represents the magnitude of the phasor corresponding to side BC in the phasor diagram triangle ABC.

5. Results

This study reveal a compelling and noteworthy exhibition of Chat GPT's adept application of the Peter Chew Method. The results obtained through this method are substantial and convincing, particularly when addressing Electrical Engineering problems that cannot be directly resolved by conventional cosine and sine rules.

In contrast, employing the conventional method proves ineffective for Chat GPT in correcting solutions for Electrical Engineering problems that are beyond the scope of direct application of cosine and sine rules, possibly due to the inherent complexity involving multiple intricate steps.

This performance gap between the two approaches underscores a significant distinction, emphasizing the pivotal role played by the Peter Chew Method in elevating the AI system's proficiency in solving complex Electrical Engineering problems to unprecedented levels.

The implications of these research findings are profound and resonant. This study underscores not only the immediate advancements achieved through the integration of the Peter Chew Method but also the overarching potential it holds for the broader landscape of AI and Electrical Engineering problem-solving solutions.

The stark contrast in outcomes serves as a testament to the transformative power that diverse methodologies can exert in shaping the capabilities of AI systems, unlocking novel avenues for exploration and comprehension in solving intricate Electrical Engineering problems.

Conclusion

Pioneering novel mathematical methodologies, such as the Peter Chew Method, has become imperative for overcoming errors in solving Electrical Engineering problems within AI systems like GPT Chat. The groundbreaking nature of the Peter Chew Method is evident in its transformative impact on the field of mathematics, specifically addressing the persistent challenge of providing correct answers for complex Electrical Engineering problems in AI systems.

This study stands as a resounding testament to the critical need for innovative rules when tackling such challenges, exemplified by the ingenuity of the Peter Chew Method. By venturing into unexplored territories and devising inventive solutions, this research illuminates the significant potential inherent in adopting novel approaches.

The Peter Chew Method, tailored to provide accurate and straightforward solutions for Electrical Engineering problems, emerges as a guiding light, offering a promising avenue for AI systems like ChatGPT to transcend errors.

This innovative rule promises to enhance the proficiency of A systems to unparalleled levels.

In essence, the Peter Chew Rule serves as a strategic gateway to expanding the horizons of AI capabilities. It embodies a fresh perspective that not only addresses errors in AI systems like Chat GPT but also charts a course toward pushing the boundaries of what artificial intelligence can achieve.

This study underscores the inherent power within inventive rules, illustrating their pivotal role in shaping the trajectory of AI advancement. As AI systems continue to evolve, embracing pioneering rules becomes not merely an option but a necessity for unlocking the full potential of artificial intelligence.

7. Discussion

The insights derived from this study distinctly emphasize the pivotal role that methodological choices play in enhancing the mathematical capabilities of ChatGPT. Among the diverse approaches explored, the application of the Peter Chew Rule emerges as a particularly noteworthy and impactful example.

The Peter Chew Rule stands as a testament to innovation in overcoming inherent challenges in AI systems like ChatGPT, specifically addressing the persistent issue of providing correct answers for solving Electrical Engineering problems that cannot be resolved directly by conventional cosine and sine rules.

By seamlessly integrating this rule, ChatGPT is empowered to transcend its previous errors and directly solve complex Electrical Engineering problems.

The Peter Chew Rule represents a profound advancement in AI's ability to navigate intricate mathematical scenarios. Its integration signifies a pivotal turning point, underscoring the significance of innovative methodologies in reshaping the landscape of AI solutions.

In essence, the utilization of the Peter Chew Rule marks a decisive step towards refining ChatGPT's mathematical prowess. By effectively bridging the gap between existing capabilities and the pursuit of more comprehensive solutions, this rule symbolizes the potential for exponential growth in AI's mathematical competencies.

As AI systems continue to evolve, methodologies like the Peter Chew Rule serve as guiding beacons, illuminating the path towards enhanced proficiency and groundbreaking achievements.

.

. Implications and Future Research:

he discoveries unveiled in this study hold the potential to evolutionize the mathematical aptitude of AI, emphasizing the nperative for continuous innovation in new rules, theorems, ethods, or formulas to propel AI systems, such as ChatGPT, to ven greater heights.

his research serves as a clarion call for future endeavors in xploring and creating innovative mathematical techniques neticulously tailored to the evolving needs of AI systems.

his avenue of exploration promises to usher in a new era of AI apabilities, extending their influence across a diverse spectrum f problem-solving domains.

The pursuit of novel methodologies could yield solutions that esonate deeply with students, thereby enhancing their ngagement with AI systems like ChatGPT—particularly elevant in the context of unforeseen challenges, such as those osed by events like the COVID-19 pandemic.

7

As we navigate the future, the symbiotic relationship betwee AI and mathematics is poised to thrive through inventiv strategies. By developing methodologies attuned to AI systems we unlock the potential for heightened student interest i leveraging tools like ChatGPT for learning and solvin, Electrical Engineering problems.

This synergy could offer a valuable resource to mitigat learning disruptions during periods of crisis, fostering dynamic and adaptable learning environment for students ever when confronted with unprecedented circumstances.

. Reference:

. Introducing ChatGPT. OpenAI .
ttps://openai.com/blog/chatgpt

. Matt G. Southern OpenAI's ChatGPT Update Brings
mproved Accuracy. Search Engine Journal. January 10, 2023.
ttps://www.searchenginejournal.com/openai-chatgpt-
pdate/476116/#close

. *James Lin ,*Knowledge is power: why the future is not just
bout the tech . World Economy Forum Jan 25, 2021
ttps://www.weforum.org/agenda/2021/01/knowledge-is-
power-why-the-future-is-not-just-about-the-tech/

. Lesson 5. PHASE RELATIONS AND VECTOR
REPRESENTATION Electrical Engineering.

ttp://ecoursesonline.iasri.res.in/mod/resource/view.php?id
=93227

5.   Phasor Diagrams and Phasor Algebra. **Electronics Tutorials**

https://www.electronics-tutorials.ws/accircuits/phasors.html

6. Chew, Peter, Peter Chew Method For Solution Of Triangle (August   29,   2022).   Available at   https://ssrn.com/abstract=4203746 or http://dx.doi.org/10.21 39/ssrn.4203746

7. Triangle Calculator - shows all steps - Math Portal https://www.mathportal.org/calculators/plane-geometry-calculators/sine-cosine-law-calculator.php

8.   Chat *GPT* . https://chat.openai.com/?model=text-davinci-002-render-sha .